SpringerBriefs in Business

SpringerBriefs present concise summaries of cutting-edge research and practical applications across a wide spectrum of fields. Featuring compact volumes of 50 to 125 pages, the series covers a range of content from professional to academic. Typical topics might include:

- A timely report of state-of-the art analytical techniques
- A bridge between new research results, as published in journal articles, and a contextual literature review
- A snapshot of a hot or emerging topic
- An in-depth case study or clinical example
- A presentation of core concepts that students must understand in order to make independent contributions

SpringerBriefs in Business showcase emerging theory, empirical research, and practical application in management, finance, entrepreneurship, marketing, operations research, and related fields, from a global author community.

Briefs are characterized by fast, global electronic dissemination, standard publishing contracts, standardized manuscript preparation and formatting guidelines, and expedited production schedules.

Uwe Wehrspohn • Dietmar Ernst

When Do I Take Which Distribution?

A Statistical Basis for Entrepreneurial Applications

 Springer

Uwe Wehrspohn
Wehrspohn GmbH & Co. KG
Mannheim, Germany

Dietmar Ernst
University of Nürtingen-Geislingen (HfWU)
Nürtingen, Germany

ISSN 2191-5482   ISSN 2191-5490   (electronic)
SpringerBriefs in Business
ISBN 978-3-031-07329-8   ISBN 978-3-031-07330-4   (eBook)
https://doi.org/10.1007/978-3-031-07330-4

This Springer imprint is published by the registered company Springer Nature Switzerland AG
The registered company address is: Gewerbestrasse 11, 6330 Cham, Switzerland

Preface

One of the most frequently asked questions in enterprise risk management is, "when do I use which distribution to represent enterprise risk"? It is asked by beginners as well as by experienced risk managers who are faced with the task of opening the chapter of quantitative risk management for the first time in the company.

There is no canonical answer to this question. It is not the case that the financial damage of a loss of production due to flooding of the plant area generally obeys a triangular or a normal distribution. The same applies to all other types of risks.

However, we can consider the properties of many important distributions and then consider which distribution with their properties fits best in a specific situation.

We can also discuss how to determine the parameters of a distribution and whether this is practically possible in the given situation. Because only if we can parameterize a chosen distribution, we can use it.

In fact, this is a central and delicate step in the survey and assessment of risks in the risk management process because many risk experts who have to assess the factual context are generally not risk managers, but are responsible for completely different tasks in the company. Nevertheless, they must be in a position to comment on these issues.

In the everyday life of companies, the determination of parameters therefore often plays the central role in the selection of distributions in Enterprise Risk Management (ERM). Practically only distributions that are easy to explain and can be parameterized with the help of expert assessments are used. Parameterizations through the statistical evaluation of data on risk play only a secondary role.

We will therefore concentrate on this group of distributions for the time being. It includes above all the constant distribution, the triangular distribution and the PERT distribution, but also the uniform distribution and the trapezoidal distribution. Many companies use them to model all risks. Less well known is the modified PERT distribution, which has a shape parameter that can be used to adjust the flattening of the long side, and custom distributions, where the user can draw the density himself.

All of these distributions are bounded on both sides. Distributions unbounded on one or both sides are less often parameterized with expert estimates. Where they are nevertheless used, the normal and lognormal distributions arc usually employed. Sometimes also the Weibull distribution. All of these are classic textbook distributions. We newly introduce the Expert and the Poly distributions, both of which can be parameterized by experts in two-sided and one-sided limited and unlimited variants.

Thus, the status quo is guided by a very pragmatic rule. 'Use distributions you know and can parameterize'. In what follows, we will portray these distributions, discuss their uses and show how they can be parameterized using expert judgments and algorithms.

In Enterprise Risk Management, the modelling of risks is divided into two steps. First, the occurrence of risks is presented, and in the second step, the impact of the risk in the event that it occurred.

We take this dichotomy as an outline and group the representation into distributions that model the entry side of a risk and distributions that are used to describe the potential loss amounts.

We thank Springer Publishers for the professional realization of this book project and wish the readers an interesting read.

Mannheim, Germany   Uwe Wehrspohn
Nürtingen, Germany   Dietmar Ernst

Contents

Chapter 1
Distributions for the Occurrence of the Risk

Distributions that model the entry side of a risk count how often the risk actually materializes in a period.

In principle, all distributions that take the counting numbers 0, 1, 2, 3, etc. as values can be used for this purpose. In practice, however, three distributions are discussed and two of them are actually used. These are the Bernoulli, Poisson, and binomial distributions, with the first two being used in practice.

1.1 Bernoulli Distribution

The Bernoulli distribution is the classic occurrence model in enterprise risk management. Its position dates back to the time when risk was represented by a probability of occurrence and an impact—two numbers each. The use case was very specifically described operational risks in a one-period view.

The Bernoulli distribution describes a generalized coin toss. There are exactly two states. An event ("head" or, transferred to our case, "risk materializes") occurs or does not occur. Multiple occurrences are not possible (Fig. 1.1).

The distribution has one parameter, the probability of occurrence **P** of the risk under consideration (Fig. 1.2).

In Enterprise Risk Management (ERM), the probability of occurrence (OP) is in most cases determined by expert assessment. This is also necessary because in ERM many risks are included in the analysis that have not yet occurred in everyday life but are considered possible.

If risk experience is available in the company, the OP can also be estimated by dividing the number of observed occurrences of the risk by the number of observation years. Thus, if the risk has occurred for example twice in the last 10 years, according to this logic the OP would be estimated at 20%.

© The Author(s), under exclusive license to Springer Nature Switzerland AG 2022
U. Wehrspohn, D. Ernst, *When Do I Take Which Distribution?*, SpringerBriefs in Business, https://doi.org/10.1007/978-3-031-07330-4_1

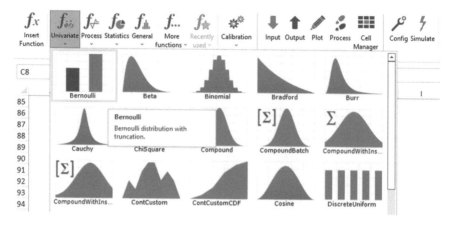

Fig. 1.1 The Bernoulli distribution on the Risk Kit toolbar

Fig. 1.2 Input dialog of the Bernoulli distribution in Risk Kit

In order to broaden the scope of risk knowledge, it is quite possible to include experience from the market. If a risk has occurred at other companies, there may be much to suggest that it could also happen at your own company.

The evaluation of data for determining the probability of occurrence is all the more reliable the more representative occurrences there are for the risk. The rarer the risk, the more difficult and error-prone it becomes. If, in the extreme case, the risk has never occurred, it is not possible to reach the goal without expert assessments.

In the simulation, the Bernoulli distribution takes the values 0 and 1. The 1 represents the occurrence of the risk (Fig. 1.3).

This property of the Bernoulli distribution, that a risk either cannot occur or can occur exactly once, but never more than once, is an essential criterion for its application.

Fig. 1.3 Bernoulli distributed random number

In many ERM models that look at the evolution of risks over multiple periods, the Bernoulli distribution creates an inconsistency in the model. In multi-period models, risk occurrence is simulated in each period. If a Bernoulli distributed risk has occurred in period 1 in a simulation run, this risk is henceforth blocked for the period and cannot occur any further. However, as soon as the next fiscal year (period 2) has begun, the risk is unlocked again and can occur again from January.

The same contradiction arises when periods are divided. If one changes from an annual to a quarterly view in a Bernoulli model, the risk can suddenly occur four times as often over the original period of one year as before. Conversely, if one switches to a 5-year view for strategic risks, for example, the risk will occur less frequently, even if the probabilities of occurrence are correctly adjusted to the changed periods.

The Bernoulli distribution is best suited as a model for the occurrence of a risk if this risk can only occur once, regardless of the delimitation of the periods.

For the original use case "specific operational risks with a one period observation horizon" this criterion is very well fulfilled. A specific product is taken off the market only once. A named bridge collapses only once, etc.

This changes in the case of more generally formulated descriptions of operational risks ("One of our products has to be taken off the market," "A supply route becomes impassable and the supply chain is interrupted"). Analogously for extended risk terms.

An extension of the modelling of risk entries, therefore, allows multiple entries of a risk in a period. The two most important tools for this are the binomial and Poisson distributions.

Mathematically, the Bernoulli distribution is the building block from which these two (and many other) distributions are constructed. It is therefore a direct generalization.

1.2   Binomial Distribution

A binomial distribution arises when we have a fixed number **n** of Bernoulli experiments in which only 0 or 1 (success or failure, risk fails, or risk occurs) can come out.

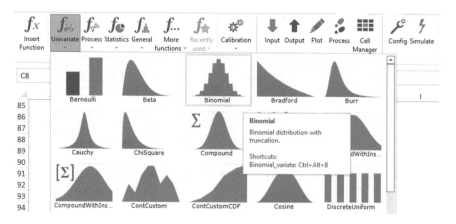

Fig. 1.4 The binomial distribution on the Risk Kit toolbar

The total number of successes on **n** trials takes a value in the count numbers 0, 1, 2, ..., n.

Example: If I operate a wind farm of 8 wind turbines and cannot easily replace one turbine, 0–8 wind turbines may fail at the same time (Fig. 1.4).

The binomial distribution has two parameters. The number of trials **n** and the probability of success in each trial **p** (Figs. 1.5 and 1.6).

An important assumption of the binomial distribution is that the probability of success remains the same for each experiment. In practice, this can be true, but it can also be a limitation.

For the wind farm in the example, the assumption of equal probabilities of occurrence for the risk of failure of a wind turbine would be well met if the turbines were very similar in model, load, and age. However, if the wind farm is a mixture of smaller and larger turbines of different types and with different operating durations, it might be more realistic to consider a separate risk for each turbine.

The binomial distribution as a model for the number of system failures would also not be ideally suited if a system only failed temporarily and went back into production after repair. In this case, it would be possible for individual wheels to fail several times and for more failures to occur in individual cases than there are plants in the park.

In a multi-period model, if the binomial distribution is used, there may be dependence between periods if the maximum frequency of occurrence in a period is changed by the number of claims in a previous period.

If, in the example, two wind turbines have permanently failed in period 1, only 6 of the original turbines will remain intact in the subsequent periods, so that the value of **n** would have to be reset here. Additional damage would further reduce the number of still intact turbines.

By breaking down the risk into one risk per plant, which is finally dismantled after failure, this case can also be simplified in the same way as above, so that the time dependencies arise by themselves.

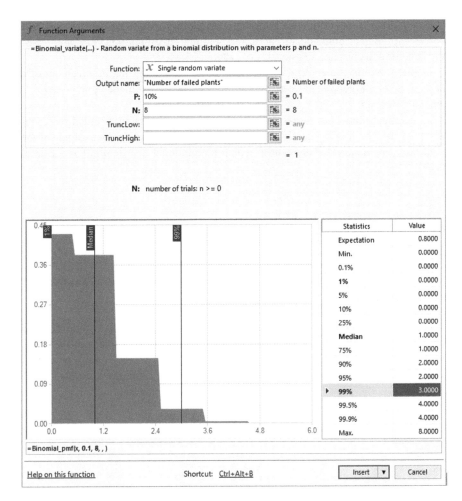

Fig. 1.5 Input dialog of the binomial distribution

Fig. 1.6 Binomially distributed random number

Due to these complexities and the general possibility of splitting the risk into risks with Bernoulli distributed occurrences, the binomial distribution is very rarely used for modelling occurrence frequencies of risks in ERM.

In ERM, due to the enterprise-wide use of the model and the involvement of a large number of people, a certain standardization of the model components is generally desired.

In technical models of large-scale plants it is different. Here, detailed representations of smaller, homogeneous groups of plants, as described in the example of the wind farm, are a standard element in which the binomial distribution has a central place.

1.3   Poisson Distribution

If an event occurs over a period of time with constant probability ("intensity") and these occurrences are independent of each other, its frequency of occurrence is Poisson distributed.

It takes values on the count numbers 0, 1, 2, ... (Fig. 1.7).

Its parameterization is done via the expected frequency of occurrence **lambda** of the risk. This is a major advantage of the distribution in the context of the ERM process, as the expected frequency of occurrence is clearly understandable to experts and can therefore be determined with good justification. It can also be obtained very

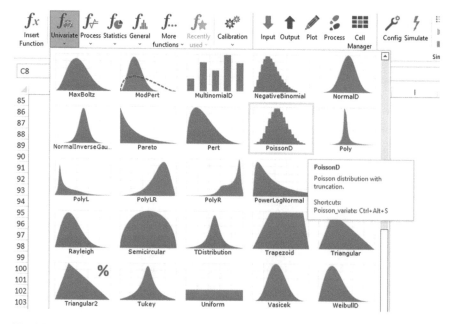

Fig. 1.7 The Poisson distribution on the Risk Kit toolbar

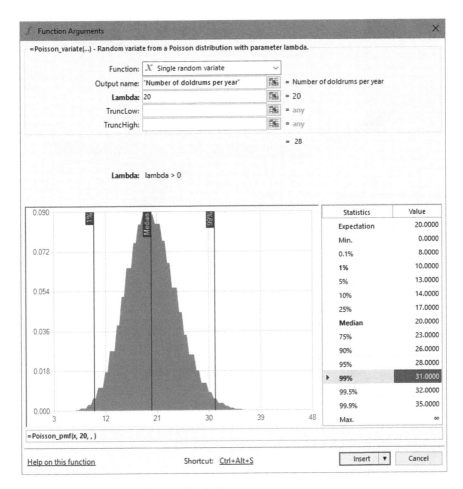

Fig. 1.8 Input dialog of the Poisson distribution

well from data as an expected value of the frequencies of occurrence, if such data are available.

Example: In the wind farm example, the number of lulls, i.e., periods of low wind of a certain minimum duration, is an important factor for the quality of the site and the profitability of the plant.

The expected number of doldrums per year can be determined from the weather data of the past years (Fig. 1.8).

In this example, an important property of the Poisson distribution is that we do not need to specify a maximum number of doldrums (Fig. 1.9).

Poisson and binomial distributions are approximately interchangeable in many cases. This is always the case when the expected occurrence frequency **lambda = n * p** is small compared to **n**. The deviations between the two distributions are then usually so small that they play no practical role in the ERM process. Both

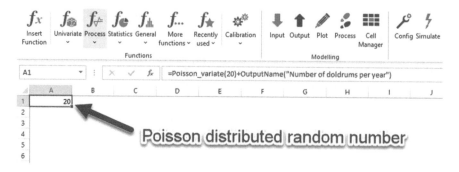

Fig. 1.9 Poisson distributed random number

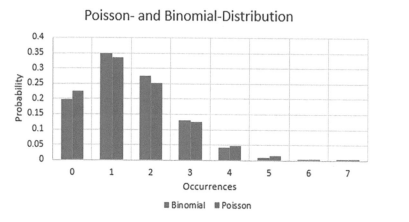

Fig. 1.10 Poisson and binomial distributions in comparison

distributions even become equal when, for a given expected occurrence frequency, **n** becomes large.

Figure 1.10 gives an example of the comparison of both distributions for **lambda** = 1.5, **p** = 15% and **n** = 10.

Because of these properties, the Poisson distribution is the most widely used model in ERM for representing frequencies.

Chapter 2
Distributions for the Effects of Risk

To fully assess a risk, after the description of the occurrence of the risk, the amount of loss after occurrence is relevant. It is generally assumed that each occurrence causes an individual loss. Thus, if a risk occurs more than once, the total loss is the sum of the individual losses. We will discuss this in detail in the section on the compound distribution, which evaluates a random number of losses of a random amount in each case.

The distribution most commonly used in companies in the past for single loss occurrence is the constant distribution. This point-based representation is often perceived as unrealistic. Attempts are therefore being made to replace it with ranges. These ranges may well include opportunities. Distributions that are often used for this purpose are the uniform, triangular, PERT, and trapezoidal distributions.

All of these distributions are directly related to the classic business best- and worst-case considerations and were originally used for estimating operational risks, i.e., operational losses. In these cases, a worst case in the sense of "tear down and rebuild" is often well defined.

The situation is different for risks whose effects cannot be so easily capped. What is the worst-case scenario in the event of a pandemic? Some companies have failed precisely at this point with a realistic assessment of the risks. After all, they have included the pandemic in the risk inventory (which most of the companies concerned have not done), but have underestimated the impact by a factor of tens. The assessment is quickly rendered moot by such an error.

Distributions often used for this application are the lognormal and the Weibull distribution. Both are extreme value distributions and can therefore potentially also assume very large values with a small probability.

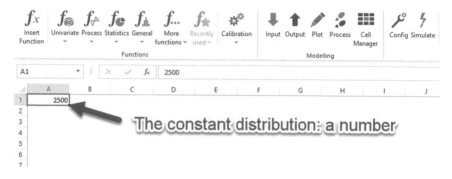

Fig. 2.1 The constant distribution in excel

2.1 Constant Distribution

Risk management textbooks of the past have described a risk by a probability of occurrence and an impact. The impact is a fixed number. This "constant distribution" is still the way things are in many companies today.

The representation of a risk by two values has advantages.

- A risk has a short and precise looking profile.
- All risks are standardized. Very different types of risk can be reported so well.
- Risk quantification seems straightforward.
- Both variables ("OP and loss amount") can be compared graphically. If necessary, categorized and underlaid with colored gradations, this results in the risk map.

The constant distribution is so simple in its technical implementation that you do not need any tools for it. It provides random numbers that you know in advance, so you do not have to draw them at all. It can be represented in Excel by a simple number (Fig. 2.1).

On closer inspection, however, describing the impact of a risk in terms of a fixed figure often turns out to be a well-meaning illusion. Only in the case of a few risks is the exact amount of the loss known in advance when it occurs. A threatened contractual penalty or the certain replacement of a wear part could be such cases.

In most contexts, however, the impact of a risk can realistically only be defined in terms of bandwidths, often a very wide range.

2.2 The Uniform Distribution

If one follows the paradigm of a range for the possible losses after the occurrence of risk, the uniform distribution corresponds to this picture in a natural way (Fig. 2.2).

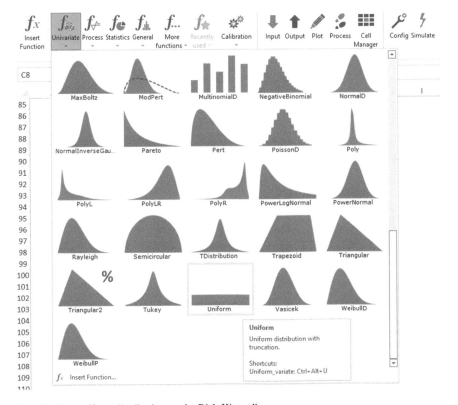

Fig. 2.2   The uniform distribution on the Risk Kit toolbar

It takes values between a minimum **A** and a maximum **B** and its density has exactly the shape of a band (Fig. 2.3).

Each value between **A** and **B** is assigned the same probability (Fig. 2.4).

Uniformly distributed random events occur in many games, often in their integer version. The sides of a die show uniformly distributed on top. Playing cards are shuffled in a uniformly distributed manner. The ball of a roulette wheel selects a digit in a uniformly distributed manner. The uniform distribution here is just the epitome of fairness.

We can harness this property in ERM.

In the wind farm example, if we had data on the length of doldrums in recent years, we can number the observed records and draw a number across the uniform distribution. The cost of the simulated doldrum is then the length of the doldrum drawn times the lost revenue per unit time.

With this approach, we avoid the need to determine the distribution of the doldrum lengths and can directly access the observed data. However, we will also never simulate a doldrum that is longer than the longest observation in our sample.

Other approximately uniformly distributed quantities include the exact location where a pipeline will leak, a rope will break, or a cable will snap.

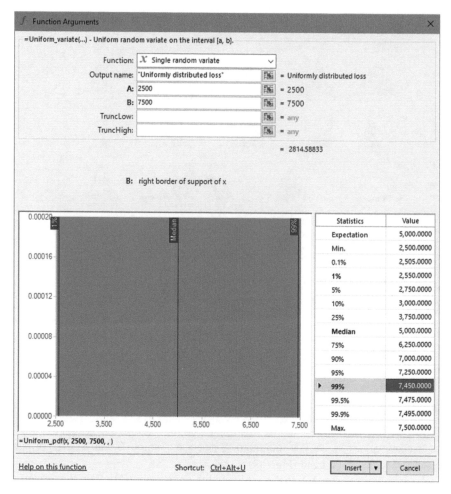

Fig. 2.3 Input dialog of the uniform distribution

Although it captures the bandwidth picture so directly, the uniform distribution in the ERM is often not immediately plausible. One important reason for this is that the density does not evolve steadily. Damage outside **A** and **B** have density 0. Values in this range are not assumed. Exactly at **A,** however, the density then jumps to a high level, remains there until **B** is reached, and then falls abruptly back to 0.

It would be more understandable to have a model in which the unadopted range merges seamlessly into the range in which the damages occur. This will give rise to the introduction of further distributions in a moment.

Precisely this disadvantage of the uniform distribution, namely that damages in the extreme ends **A** and **B** of the value range are assumed with an unrealistically high probability, is paradoxically a reason why the uniform distribution is often used, at least transitionally.

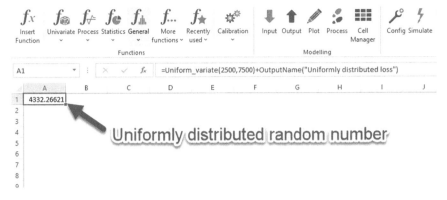

Fig. 2.4 Uniformly distributed random number

On the one hand, the choice of distribution may signal that the exact shape of the frequency trend in the damaged area is either not known or has not been investigated. Something like a most likely value or the more precise shape of the distribution is then not known. Here, therefore, an approximate solution is made clear (Pierre-Simon Laplace's indifference principle).

On the other hand, the uniform distribution fluctuates more strongly on the range of values than the alternatives triangle, PERT, or trapezoid. The overall risk, therefore, becomes more sensitive to this risk factor than if another model is chosen. Therefore, if a risk does not have a large effect on the outcome when the uniform distribution is chosen, this will not change much even when the effects are more refined. The choice of extreme values **A** and **B** is then initially decisive. If they are correct, a more differentiated representation of the impact distribution may not be worthwhile and the above-mentioned indifference principle can justifiably be applied.

2.3   Triangular, PERT, and Modified PERT Distribution

The triangular and PERT distributions are an important alternative to the uniform distribution when designing the range of possible losses from a risk. Here, the loss experience is structured over three points, the minimum, the mode, and the maximum.

The mode is the high point of the density of the distribution. It is also referred to as the "most likely value" because it is the most likely to observe damage in an environment around this point.

The determination of these key data by means of the questions "In which range do the losses lie minimally?," "In which range do the losses lie maximally?" and "In which range do we most likely expect to see losses?" is very easily possible even for risk experts who do not normally deal with the parameterization of probability

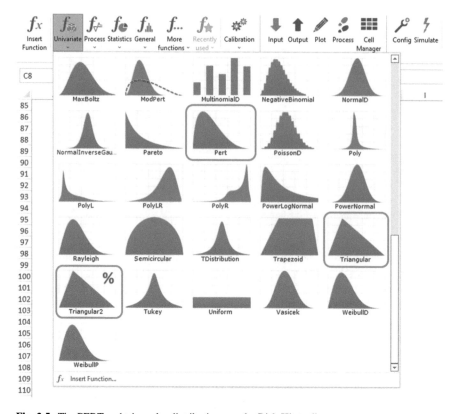

Fig. 2.5 The PERT and triangular distributions on the Risk Kit toolbar

distributions in everyday life. This suggestive power is so strong that many compa-nies base their entire ERM on these two loss distributions alone (Fig. 2.5).

Figure 2.6 shows the input dialog for the PERT and the uniform distribution. The same values are entered so that the distributions can be compared.

While both density functions take the values 0 and their maximum at the same points, the PERT distribution falls off faster at the long end than the triangular distribution. Thus, the maximum damage is assumed less frequently when using the PERT distribution in this example than with the triangular distribution. With the PERT distribution, on the other hand, more probability mass is close to the most likely value.

This difference is often a helpful criterion for determining which distribution fits best in a situation.

This consideration of including the rate at which the distribution decays toward the extreme values in the selection of the distribution has led to the formulation of the modified PERT distribution (Fig. 2.7).

The modified PERT distribution is also determined at its core by minimum, most likely value and maximum, but contains an additional parameter that determines the curvature of the distribution (Fig. 2.8).

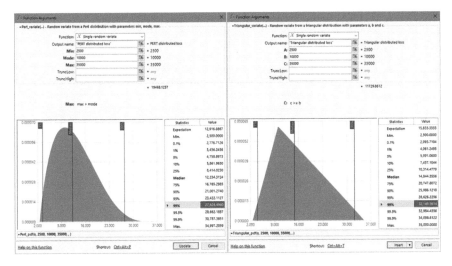

Fig. 2.6 Input dialog of the PERT and the triangular distribution

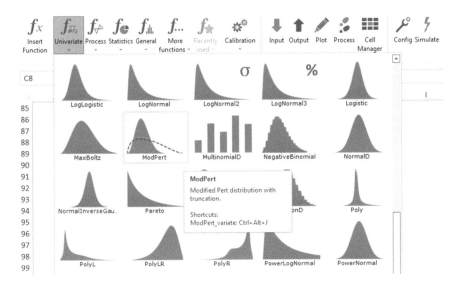

Fig. 2.7 Modified PERT distribution on the Risk Kit toolbar

For a value of the shape parameter of 4, the familiar PERT distribution is again obtained. For a value of the shape parameter <4 more probability goes to the extremes and for a value >4 the probability mass concentrates around the most likely value (Fig. 2.9).

The modified PERT distribution thus spans the entire spectrum from a uniform distribution to a point mass at the most probable value.

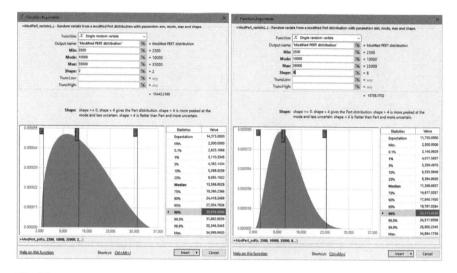

Fig. 2.8 Input dialog of the modified PERT distribution

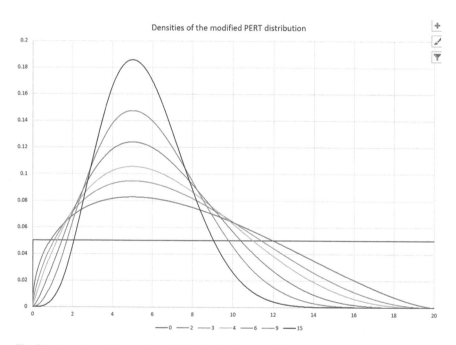

Fig. 2.9 Densities of the modified PERT distribution for different shape parameters

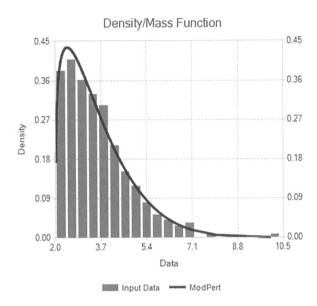

Fig. 2.10 Fitting the modified PERT distribution to measured slack lengths in days

Because of this flexibility, the modified PERT distribution can not only be well estimated by experts, but also fits observed data very well when such data are available.

In the wind farm example, for example, data on the length of lulls are available. The supplied measured values are filtered for a minimum length of 2 days (Fig. 2.10).

Although they are suggestive values, minimum and maximum are not easy to determine for experts in some situations because it is not always possible to imagine what can go wrong.

By comparison, it is often easier to retreat to the range of represented experience and leave the extension of the range of values to extreme events to the distribution.

As a first step to follow this idea, there is a second parameterization of the triangular distribution, where the most probable value is given and a range around it, which is not left with a given probability. The distribution is then drawn out on the sides to form a full triangular distribution (Fig. 2.11).

In this example, 90% of the probabilities are between 5000 and 25,000. The distribution thus takes on values between approx. 3000 and 32,000 in total.

Altogether, these function calls of the distributions result (Fig. 2.12).

2.4 Trapezoidal Distribution

A variant of the triangular distribution is the trapezoidal distribution. In this case, there is not a most probable value, but a most probable range of values.

The trapezoidal distribution thus opens up more scope for describing an "everyday case" for the losses that occur. One does not then have to commit oneself to a

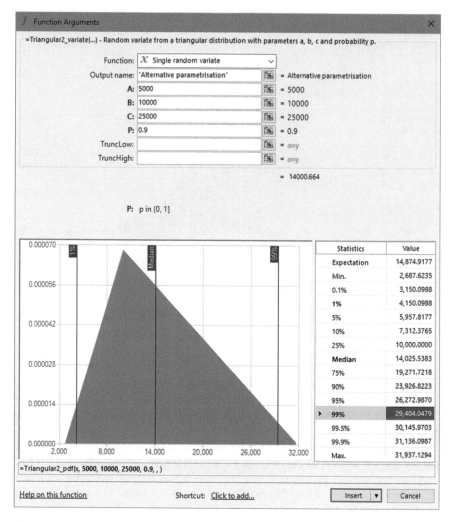

Fig. 2.11 Alternative parameterization of the triangular distribution

specific value as the most likely value, but can argue that a certain range between the extremes of the entire bandwidth has the highest representativeness for the cases of damage that one sees in connection with the risk in practice.

The trapezoidal distribution thus combines the advantages of the triangular and the uniform distribution (Fig. 2.13).

To mark out the most probable range of values, the trapezoidal distribution therefore has a fourth parameter: minimum, beginning of most likely range, end of most likely range, maximum (Fig. 2.14).

In this example, losses between 5000 and 15,000 are the most likely range (Fig. 2.15).

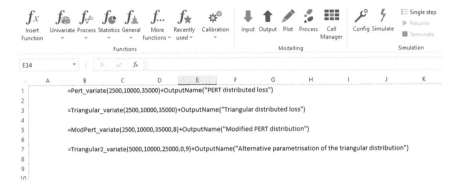

Fig. 2.12   Calling random numbers

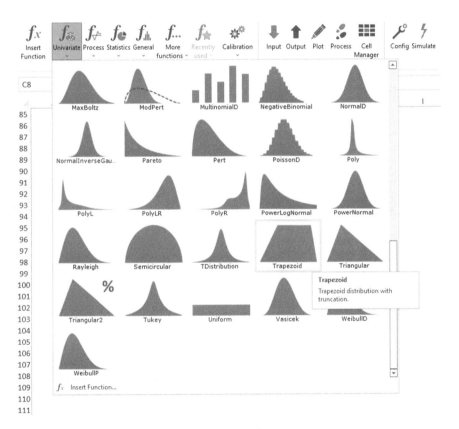

Fig. 2.13   Trapezoidal distribution on the Risk Kit toolbar

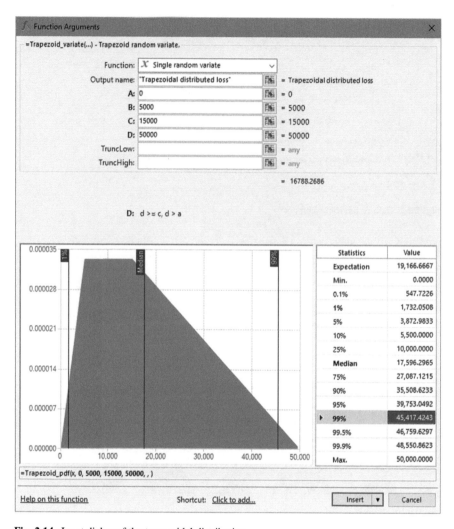

Fig. 2.14 Input dialog of the trapezoidal distribution

Fig. 2.15 Trapezoidal distributed random number

2.5   Custom Distributions

Another variation of the triangular distribution are the custom distributions. In this case, the course of the loss can be drawn individually. Special features of the loss experience, for example, in the case of IT risks, can be mapped appropriately in this way, which would generally not be possible with the triangular distribution alone (Fig. 2.16).

Cyberattacks are a common occurrence in many organizations. However, the damage progressions are a challenge for the representation in ERM. They effectively split into two regimes. Most of the time, the losses are very small and can be well represented using the triangular distribution in a narrow range. In rare cases, however, the same risk has far more dramatic effects and takes on large values with a small probability.

With the individually definable distribution ContCustom, the mapping of this risk is possible in a natural way (Fig. 2.17). The density can be drawn point by point. Thus, we can extend the usual triangle to the right to include the large losses.

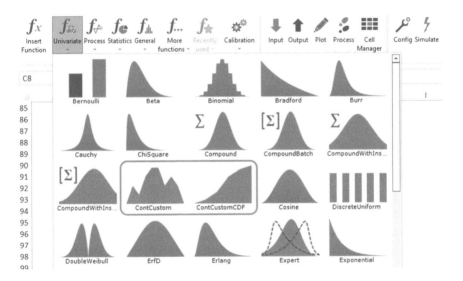

Fig. 2.16   Custom distributions on the Risk Kit toolbar

Fig. 2.17   Point-by-point
description of density

Loss	Density
0	0
5000	10
10000	0.5
100000	0

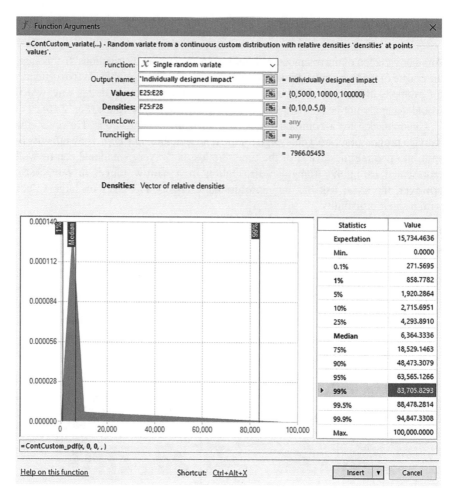

Fig. 2.18 Input dialog of the custom distribution

Risk Kit scales the values of the density so that they have the necessary mathematical properties, for example, that the area under the curve represents 100%. We can therefore confine ourselves to describing the proportions of the curve (Figs. 2.18 and 2.19).

The second custom distribution (ContCustomCDF) is conceptually structured in the same way. Instead of the density function, the distribution function is drawn from 0 to 1 point by point.

All the loss distributions discussed so far have in common that their range of values is fixed. A parameter for the maximum is explicitly specified. No loss is ever simulated above the maximum.

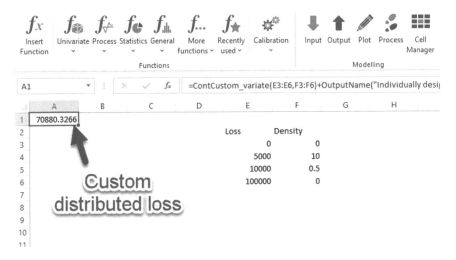

Fig. 2.19 Custom-distributed random number

The specification of the maximum is therefore critical information for all these distributions. If it is incorrectly specified, it is precisely the large losses that are systematically valued too small or too large.

In cases where the maximum loss is not precisely known, it may therefore be helpful to use distributions that are "open to the top." In this context, the normal, lognormal, and Weibull distributions are often used.

2.6   Normal Distribution

The normal distribution is certainly the best-known distribution in the general population. It owes its prominence (in Germany) to the former 10DM note on which it was depicted together with the famous mathematician Carl Friedrich Gauss, who gave a formal definition of the distribution in 1809 (Fig. 2.20).

The normal distribution is indeed encountered more often in everyday life, hence the term "normal" in the name. Many measured values fluctuate in a bell-shaped manner around a mean value (Fig. 2.21).

Moreover, sums of random variables are approximately normally distributed under very general conditions (central limit theorem). It is therefore not surprising that in many situations—and also in the ERM—we see bell-shaped frequency curves (Fig. 2.22).

An important property of the normal distribution is its symmetry around the expected value and its bell shape. It is thus clearly limited in its adaptability to special forms of impact. Its shape is quasi fixed.

In theory, the normal distribution takes values along the entire real axis. This makes it look at first as if it is well suited to capture large risk effects. However, this

Fig. 2.20 The normal distribution on the DM10 banknote

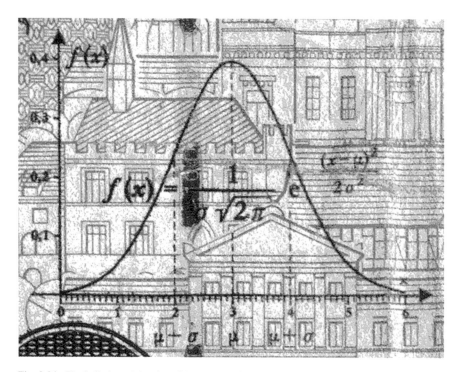

Fig. 2.21 The bell-shaped density of the normal distribution

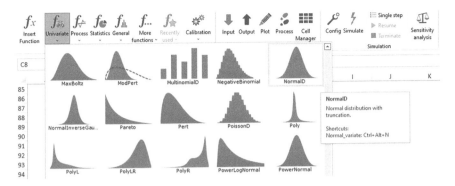

Fig. 2.22 The normal distribution on the Risk Kit toolbar

does not work in practice because its density approaches 0 so quickly that deviations from the expected value by more than 4–5 standard deviations occur so rarely that they play virtually no role for most risk measures, if they occurred at all in the simulation.

The normal distribution is thus an example of a distribution that looks unconstrained but is not. Combined with the rigidity of the shape, this is one reason why it is used in ERM only for very specific risks, notably price change risks of interest rates, exchange rates, commodities, and other market factors.

Figures 2.23 and 2.24 give an example of the use of the normal distribution to describe oil price fluctuations. This application is very comprehensively covered in Risk Kit, as with the Risk Kit Data extension you even have the possibility to load, prepare, and evaluate market data directly from the ECB and other institutions.[1]

2.7 Lognormal Distribution

A variant of the normal distribution that is far more suitable for describing exceptionally large losses is the lognormal distribution.

The lognormal distribution, starting at 0, takes values only on the positive real axis, but then may go very far up (Fig. 2.25).

It embodies the case that with the risk usually only a small to medium loss occurs and more rarely a large to a very large loss.

There are three parameterizations to choose from for the lognormal distribution in Risk Kit. First, the textbook parameterization (LogNormal), where the expected value and standard deviation of the underlying normal distribution are specified. This representation is very difficult for users to control because the transformation

---

[1] See an easy-to-read account in Wehrspohn, Zhilyakov, "Live Access to Big International Data Sources with Risk Kit Data—Case Study: Impact of Oil Price and Foreign Exchange and Interest Rates on Profit and Loss," 2021, https://papers.ssrn.com/sol3/papers.cfm?abstract_id=3824787

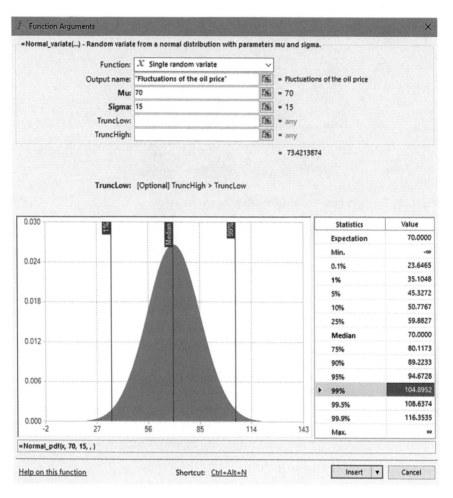

Fig. 2.23   Input dialog of the normal distribution

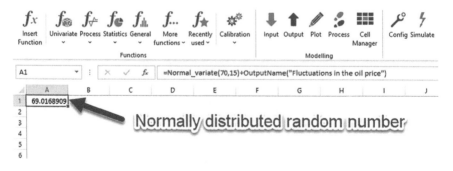

Fig. 2.24   Normally distributed random number

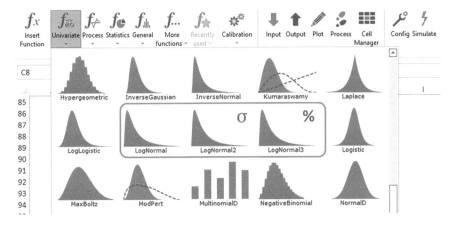

Fig. 2.25 The lognormal distribution on the Risk Kit toolbar

from normal to lognormal involves the exponential function, so the values easily explode in the simulation. This parameterization fits best when the distribution is calibrated to data.

LogNormal2 allows to directly specify the expected value and the standard deviation of the lognormal distribution. These values have the same unit as the random numbers. Thus, one has a lot of control over the magnitude one specifies. The difficulty remains that experts have little idea of what a standard deviation is, so there remains an imponderability.

Finally, LogNormal3 defines the distribution by its expected value and a quantile. Experts can therefore focus on data from their field of experience if they define the distribution precisely (Fig. 2.26).

The above example shows that large values are within the realm of possibility with the lognormal distribution. With an expectation of 25,000 and a 99% quantile of 100,000, values up to 300,000 are something that will be seen in individual cases in the simulation.

This visual feedback is very advantageous when working with the lognormal distribution, as it shows you the range of values associated with the parameterization you have chosen yourself.

Risk Kit offers a feature that additionally allows you to remove loss values that would go beyond any realistic scope—truncation. The optional truncation down or up draws a boundary into the distribution that will not be exceeded in the simulation.

In the example below, the upper limit has been set at 250,000. The loss range above this is therefore no longer relevant (Figs. 2.27 and 2.28).

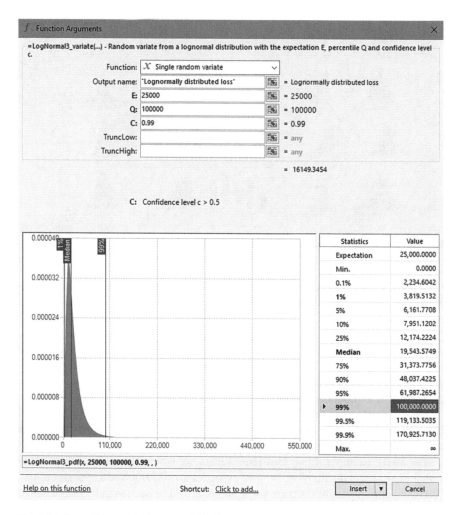

Fig. 2.26 Input dialog of the lognormal distribution

2.8 Weibull Distribution

The Weibull distribution can be used in a similar way to the lognormal distribution. It too takes only positive values and leaves room for large losses (Fig. 2.29).

Risk Kit offers two parameterizations for the Weibull distribution. One is the textbook representation (WeibullD). **A** is a location parameter that can be used to shift the distribution left and right, and is also the minimum of the distribution. **B** and **C** are shape parameters that cannot be determined via expert estimates. This parameterization is therefore only suitable for calibration on data.

The second parameterization (WeibullP) uses two quantiles in addition to the same location parameter **A** to describe the distribution. We can thus determine the

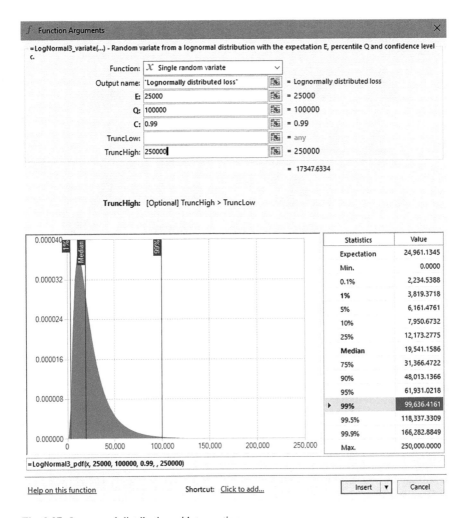

Fig. 2.27 Lognormal distribution with truncation

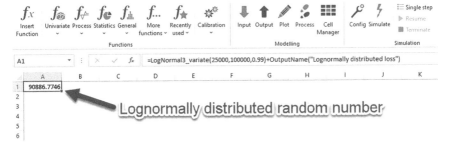

Fig. 2.28 Lognormally distributed random number

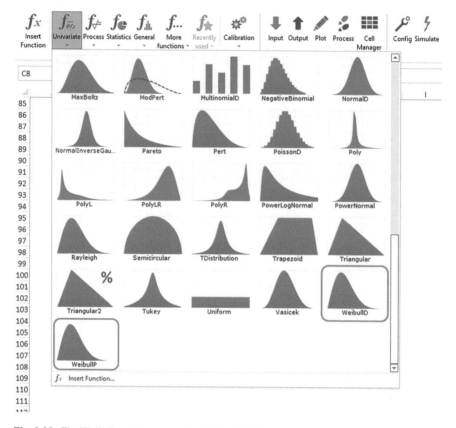

Fig. 2.29 The Weibull distribution on the Risk Kit toolbar

distribution by specifying its minimum as well as two further limits, each of which is undershot with a given probability.

For example, we can assume that losses will be positive (**A** = 0) and will remain below 35,000 with 75% probability and below 100,000 with 99% probability (**P1** = 75%, **Q1** = 35,000, **P2** = 99%, **Q2** = 100,000).

The upper tail can then go beyond this. Here, too, a visual check of the practically relevant range of values of the distribution and, if necessary, a truncation at the top makes sense (Figs. 2.30 and 2.31).

2.9 Expert Distributions

The expert distribution is a special case of the poly distributions. It determines its shape via three quantiles and, if necessary, minimum and maximum, if the distribution is to be restricted. It is thus linked to the triangular and PERT distributions, but

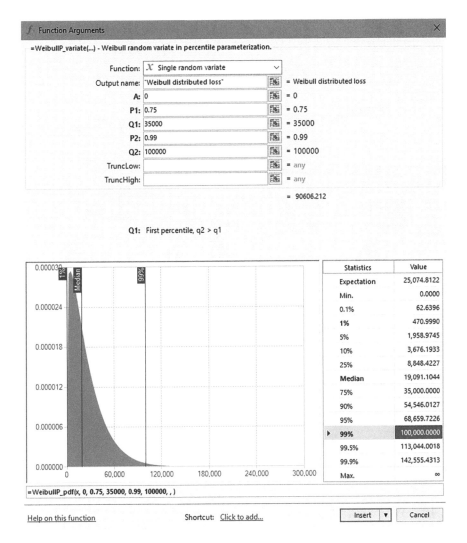

Fig. 2.30 Weibull distribution input dialog box

its shape is more flexible. For example, it also allows a one-sided or two-sided unlimited representation of effects. It can thus be used in areas where it is not possible to specify a maximum damage without further ado (Fig. 2.32).

The quantiles used to determine the expert distribution are symmetrical about the median, i.e., they are the p-quantile, the median itself and the 1-p-quantile. In addition, lower and/or upper limits for the range of values can be specified if necessary (Fig. 2.33).

The distribution goes exactly through the given points (Figs. 2.34 and 2.35).

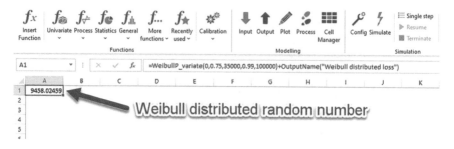

Fig. 2.31 Weibull distributed random number

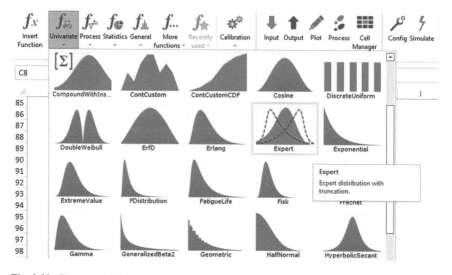

Fig. 2.32 The expert distribution on the Risk Kit toolbar

2.10 Poly Distributions

Poly distributions are a new[2] and very flexible tool in the description of risks. They gain their adaptability through their structure as a polynomial. Similar to how a Taylor series of a sufficiently high degree can approximate continuous and smooth functions with arbitrary precision, poly distributions adapt to data or expert assessments in great detail.[3]

---

[2]They were introduced by Thomas Keelin under the name MetaLog distributions. Cf. Thomas W. Keelin (2016) The Metalog Distributions. Decision Analysis 13(4):243–277.

[3]The prerequisite is that the underlying relationship is continuous in the first place and has finite moments. Expected value, variance, etc. should therefore exist and there should be no jumps or spikes.

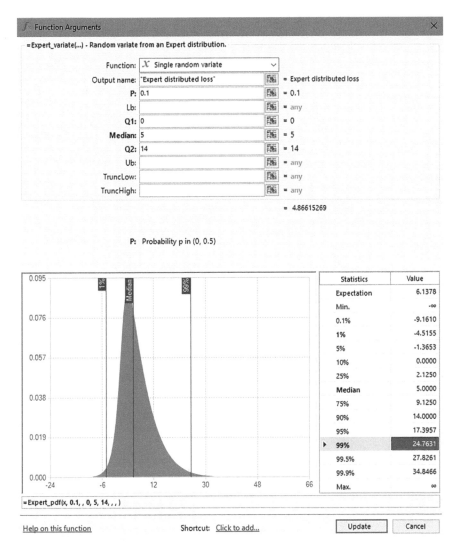

Fig. 2.33 Input dialogue of the expert distribution

Poly distributions can take all forms at the boundaries. They can be unlimited on both sides (Poly), limited on the left or right side (PolyL or PolyR) or limited on both sides (PolyLR) (Figs. 2.36 and 2.37).

The coefficients of the polynomials are abstract and not accessible to experts. If the distributions are parameterized via expert estimates, this is therefore done via a list of quantiles, i.e., pairs of values and probabilities in the *CalibratePoly* function, which determines the polynomial coefficients from the quantiles.

Unlike the expert distribution, a special case of the poly distribution, the quantiles do not have to be symmetrical about the median, but can be freely chosen. The list

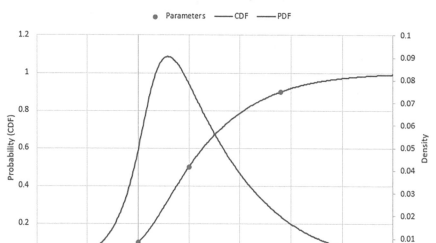

Fig. 2.34 Expert distribution and parameter

Fig. 2.35 Expert distributed random number

must contain at least three and can contain any number of points. In addition, lower and upper limits can be specified if required. The distribution is selected so that it fits the data as seamlessly as possible. Boundaries are always hit exactly (Figs. 2.38, 2.39, and 2.40).[4]

The second important application of poly distributions is their fit to data. With observed data, this is a given. But simulated data also play an important role.

---

[4]The maximum order (*maximumOrder*) can be controlled to avoid possible overfitting.

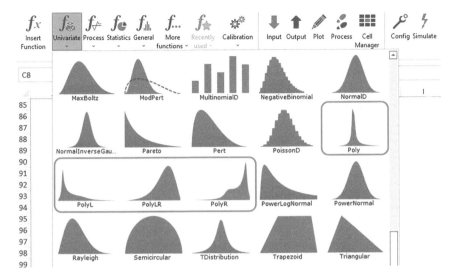

Fig. 2.36 The poly distributions on the Risk Kit toolbar

In ERM, all major risks of the company should be considered together in one analysis. In many companies, however, there are special types of risk that are so important that they receive separate attention and systems specialized in this type of risk are used that perform simulations themselves. Typical examples are treasury applications or commodity purchasing.

With poly distributions, the results of specialized simulations can be transferred to ERM very accurately with very little effort. This is also the case if the simulated distributions have a shape that cannot be mapped with standard distributions, e.g., if they have multiple modes. In simulated distributions, multiple modes arise naturally when large risks or crises occur in the simulation (Fig. 2.41).

Poly distributions are a universal key. Their flexibility allows them to bring out special features in the data that other distributions smooth out.

They can also conform to many standard distributions so precisely that for all practical purposes they can no longer be distinguished (Fig. 2.42).

2.11   Automatic Calibration of the Distributions

For all the distributions presented, there are parameterizations that can be performed by experts. This is an important prerequisite for their use in ERM, because special situations often have to be assessed.

In principle, for all the distributions mentioned, there is also the option of having Risk Kit both select the distribution that best fits the data and determine the associated parameters, if there is a data basis for a risk that can be evaluated.

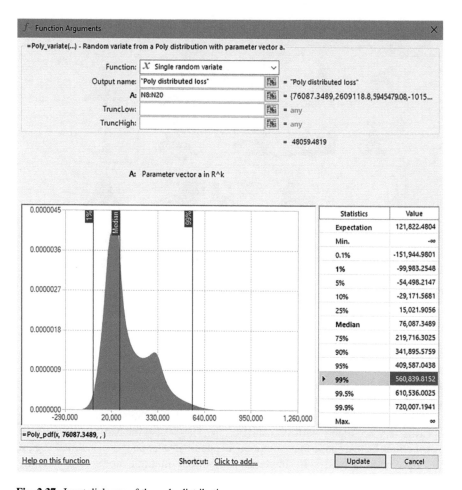

Fig. 2.37 Input dialogue of the poly distribution

Fig. 2.38 Expert
assessments of quantiles

	A	B
1		
2	Data	Probabilities
3	1	0.1
4	3	0.3
5	5	0.5
6	10	0.8
7	13	0.9
8		

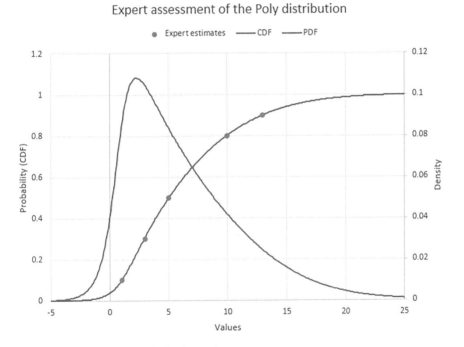

Fig. 2.39 Determining the poly distributions with the CalibratePoly function

Fig. 2.40 Adjustment of the distribution to the expert assessment

Data is routinely available in many risk areas, such as weather and climate risks, cyber risks, market risks including commodity purchasing. Thus, it may be worth looking to see if a data basis can be identified for the general situation in which the risk is embedded (Fig. 2.43).

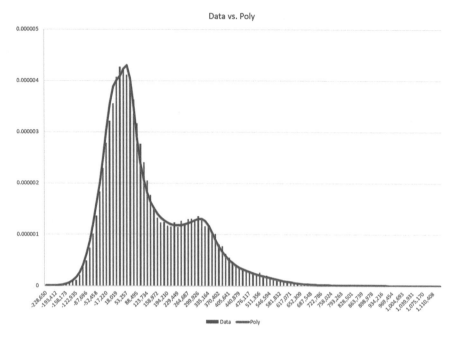

Fig. 2.41 Poly distribution fitted to simulated data

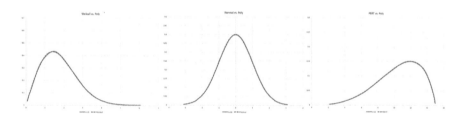

Fig. 2.42 Superpositions of Weibull, normal, and PERT distribution by poly distributions

Fig. 2.43 Calibration on the Risk Kit toolbar

You can call up the calibration dialog for one- or multi-dimensional distributions via the toolbar. There you first refer to your data basis and then select the distributions that you want to include in the analysis (Fig. 2.44).

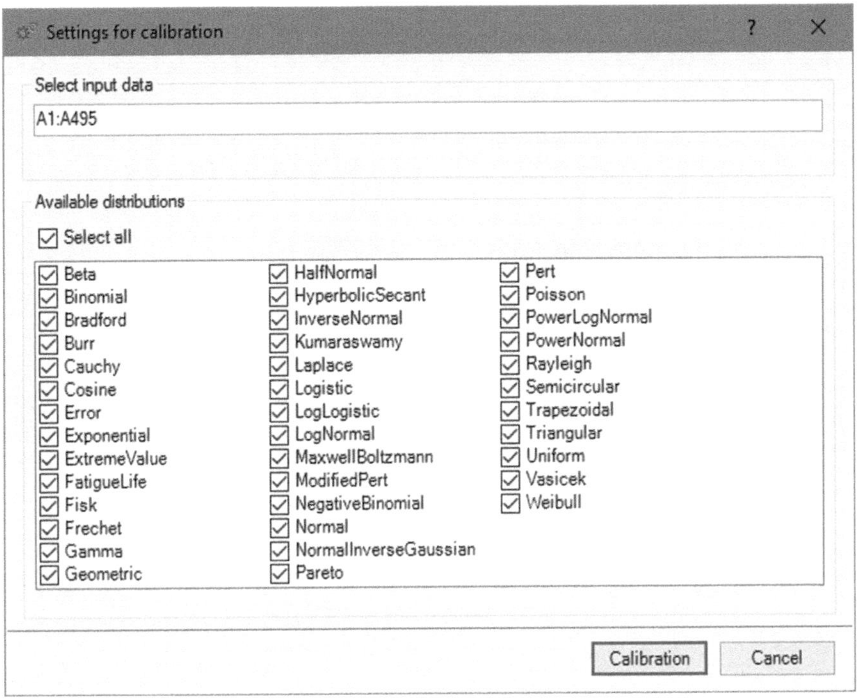

Fig. 2.44 Calibration dialog

Two strategies are recommended for selecting distributions. First, you can allow exactly those distributions that you know well and where you know what you will get. In general, it is recommended to work only with distributions that you know.

On the other hand, you can also select all distributions to see if perhaps a distribution you personally do not know fits your data very well and you might want to get to know this distribution better.

If you have selected distributions that do not match the available data, these distributions will be removed from the list and you will receive a corresponding message (Fig. 2.45).

Finally, you get the calibration result with an exact representation of the goodness of fit of each distribution to the data (Fig. 2.46). You see the fit visualized and in the form of distance measures as numbers. For each metric, the smaller the value, the better the fit. The list is sorted in ascending order.

The universe of distributions in Risk Kit is so broad overall that there are often multiple distributions that fit a given data set well. This allows you to assert, in addition to mathematical reasons for fitting, how well your organization is familiar and comfortable with the distribution, whether the distribution and its parameters are easily communicable, whether the range of values in the data is maintained or expanded, etc.

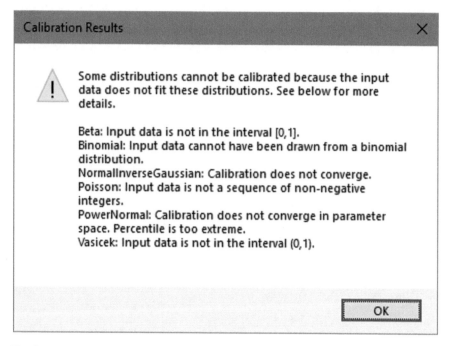

Fig. 2.45 Distributions incompatible with the data

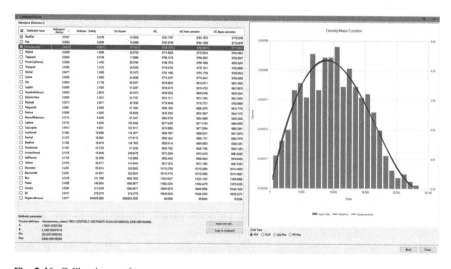

Fig. 2.46 Calibration result

You can also use calibration in combination with expert judgement, for example by pre-calibrating the distributions on the data and then reasoning further on this basis.

It may be, for example, that the data do not include certain major losses and thus paint an overly optimistic picture of the situation. Experts know this and can correct it.

2.12 Compound Distribution

The loss distributions discussed so far indicate the loss if the risk occurs once. In the case of multiple occurrences, a loss must be simulated for each individual occurrence and these values are added together to obtain the total loss from the risk.

Solving this with Excel involves additional and quite considerable effort, since the number of occurrences can be different in each simulation run. To simplify this and make it manageable, there is the compound distribution, which takes over the evaluation of a risk for the combination of occurrence frequency and loss distribution and performs the loss aggregation for this risk (Fig. 2.47).

The compound distribution combines two distributions, hence the name (Fig. 2.48). The frequency distribution is evaluated once in a simulation run. Then the frequency of risk occurrence is known.

The impact distribution may be evaluated by Risk Kit several times in one simulation run. It must therefore be specified in a form that makes this possible, namely as the name of the function followed by its parameters. The number and meaning of the parameters correspond to the selected impact distribution (Fig. 2.49).

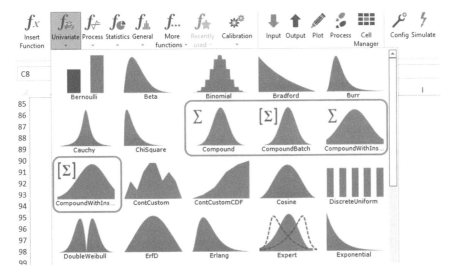

Fig. 2.47 Compound distribution on the Risk Kit toolbar

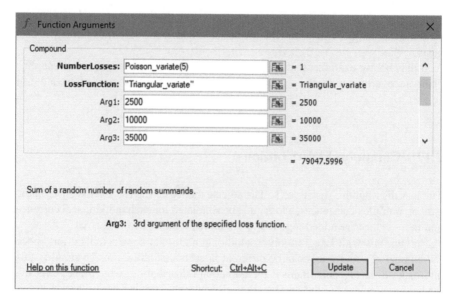

Fig. 2.48 Input dialog of the compound distribution

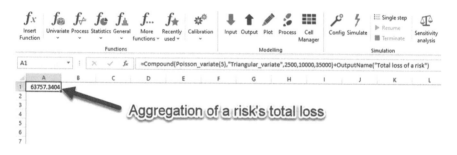

Fig. 2.49 Loss aggregation with the compound function

Additional complexity arises if the damage from the risk is covered by insurance. Maximum limits and deductibles here can relate to the individual claim and/or to the sum of claims per year.

This question is covered by the supplementary function "CompoundWithInsurance," which takes the properties of the insurance as additional values.

Chapter 3
Application of Distributions in ERM

All these distributions can be used directly with Risk Kit in risk aggregations in ERM and for PS 340. A template model that structures the application and adds an evaluation can be found on the Risk Kit toolbar under "Examples—Case Studies." The model is presented in detail in a *film*. Small and medium-sized companies will already be able to cover the requirements of PS 340 with this model.

Large companies, whose risk management process includes many locations and usually even country subsidiaries, need a more comprehensive platform to efficiently support risk collection and assessment, action management, risk analysis, and reporting.

Enterprise Risk Evaluator provides you with a modern implementation of an end-to-end quantitative risk management process. And it is designed to allow risk experts to participate in the process of risk evaluation who are not otherwise professional risk managers in their professional lives. Contact us to get to know the Enterprise Risk Evaluator in a presentation.

In seminars, we go into greater depth on the use of Risk Kit, the Monte Carlo simulation method, and its application in risk management. The topics and questions of the participants can also be addressed here.

© The Author(s), under exclusive license to Springer Nature Switzerland AG 2022
U. Wehrspohn, D. Ernst, *When Do I Take Which Distribution?*, SpringerBriefs in Business, https://doi.org/10.1007/978-3-031-07330-4_3

Printed by Printforce, the Netherlands